NUREG-1830,
Volume 9
February 2013

FY 2012   USNRC | OFFICE OF INVESTIGATIONS

AVAILABILITY OF REFERENCE MATERIALS
IN NRC PUBLICATIONS

ABSTRACT

This report provides the Commission with an overview of the U.S. Nuclear Regulatory Commission (NRC) Office of Investigations' (OI) activities, mission and purpose, along with the framework of case inventory with highlights of significant cases that the NRC OI completed during Fiscal Year 2012 (reference SRM COMJC-89-8, dated June 30, 1989). This is the 24th OI Annual Report.

TABLE OF CONTENTS

FISCAL YEAR 2012 HIGHLIGHTS

During Fiscal Year (FY) 2012, the U.S. Nuclear Regulatory Commission's (NRC) Office of Investigations (OI) sustained a mission-driven, high-performing, results-focused workforce, which enhanced its dedication to investigative excellence, effective communication, and stakeholder outreach. OI is comprised of experienced Federal criminal investigators and professional support staff who are continuously motivated to exceed the expectations of both internal and external stakeholders, while increasing opportunities for program improvement, operational awareness, engagement, and professional and technical development in accomplishing OI's role within the mission of the NRC.

The following are significant achievements during FY 2012:

- OI closed 133 investigations. Of these investigations, 98 percent developed sufficient information to reach a conclusion of substantiated or unsubstantiated regarding willful wrongdoing. This exceeded OI's performance goal of 90 percent.

- Of the 131 investigations closed with sufficient information to reach a conclusion (substantiated or unsubstantiated) related to willful wrongdoing, OI closed 85 percent in 9 months or less. OI's performance exceeded the goal of 80 percent for reactor investigations and met the goal of 85 percent for materials investigations.

- Of the 45 Assists to NRC Staff closed, OI completed 100 percent within 90 days, which exceeded OI's performance goal of 90 percent.

- OI processed 51 actions resulting from Freedom of Information Act (FOIA) requests during FY 2012 in a timely manner.

- OI referred 100 percent of its substantiated wrongdoing investigations to the U.S. Department of Justice for prosecution consideration.

- The Director of OI, representing the NRC, signed a partnership agreement with the National Intellectual Property Rights Coordination (IPR) Center. The IPR Center is a part of the U.S. Department of Homeland Security Immigration and Customs Enforcement. The agreement outlines the collaborative investigative efforts and cooperation protocols the two agencies will share related to counterfeit, fraudulent, and suspect items and equipment, including those used in nuclear power plants and devices using nuclear materials. OI also joined Operation Chain Reaction, which focuses on protecting the Nation's critical infrastructure from the introduction of counterfeit, fraudulent, and suspect items. OI accomplished these proactive efforts to support the NRC's Counterfeit Fraudulent and Suspect Items initiative in cooperation with NRC Office of New Reactor (NRO).

- OI launched its new NRC internal Web page with an added communication tool, "Ask OI Management." This communication tool is an initiative to enhance the more open, collaborative working environment within OI, along with improving communication between OI Headquarters and the OI field offices co-located with the four NRC regional offices.

- OI special agents conducted law enforcement liaison and coordination with Federal, State, and local law enforcement officials and at various State Fusion centers throughout the United States to support the NRC Federal Security Coordinator Program, as required by the Energy Policy Act of 2005.

- OI effectively addressed the emerging issues of potential violations of NRC regulations related to export of licensed materials to embargoed countries.

- OI participated in U.S. Department of Justice Anti-Terrorism Advisory Council meetings related to national security concerns and counterterrorism.

INTRODUCTION AND OVERVIEW

Mission and Authority

As stated in the U.S. Nuclear Regulatory Commission's (NRC's) Strategic Plan for fiscal years (FYs) 2008–2013, the agency's mission is to license and regulate the Nation's civilian use of byproduct, source, and special nuclear materials to ensure adequate protection of public health and safety, promote the common defense and security, and protect the environment. The NRC's vision is excellence in regulating the safe and secure use and management of radioactive materials for the public good. The mission and vision provide the framework for the agency's strategies and goals, which guide the allocation of resources across the agency.

The Office of Investigations (OI) aligns with the agency's regulatory programs and strategic values and goals to provide for the safe use of radioactive materials and nuclear fuels for beneficial civilian purposes. OI's investigations program is consistent with the agency's adherence to the principles of good regulation-independence, openness, efficiency, clarity and reliability, and by providing regulatory actions that are effective, realistic, and timely.

The Commission has delegated to the Director of OI the authority to take the necessary steps to accomplish the OI mission, as described in Title 10 of the *Code of Federal Regulations* (10 CFR) 1.36, "Office of Investigations." See Section 161(c) of the Atomic Energy Act of 1954, as amended (42 U.S.C. 2201 (c)); and Section 206 of the Energy Reorganization Act of 1974 (42 U.S.C. 5846). OI investigative jurisdiction extends to the investigation of alleged wrongdoing by licensees, certificate holders, permittees, or applicants; by contractors, subcontractors, and vendors of such entities; and by management, supervisory, and other employed personnel of such entities who may have committed violations of the Atomic Energy Act, the Energy Reorganization Act; and rules, orders, and license conditions that the Commission issued.

Additionally, during the course of a investigations, OI may uncover potentially safety-significant issues that may or may not be related to wrongdoing. In these instances, OI provides this information to the technical staff in a timely manner for appropriate action. OI also provides professional investigative support to the NRC staff when requested in the form of Assists to NRC Staff. Generally, these "Assists to Staff" are associated with matters of regulatory concern for which the staff has requested OI's investigative expertise, but that do not initially involve a specific indication of wrongdoing.

THE OFFICE OF INVESTIGATIONS

The Director of the Office of Investigations (OI) reports to the Deputy Executive Director for Materials, Waste, Research, State, Tribal, and Compliance Programs (DEDMRT) and provides investigative support to Operating Reactors, New Reactors, International Programs, and Nuclear Materials Users programs.

OI is an independent, national investigations program, which consists of four regionally based field offices headed by Special Agents In Charge (SAICs), who report directly to OI senior management located at OI Headquarters. OI field offices and headquarters are staffed by Special Agents (Job Series GG-1811 Federal Criminal Investigators) and professional support staff.

All NRC OI special agents have extensive backgrounds and experience in Federal criminal investigations. During FY 2012, the professional cadre of OI special agents possessed an average of 20 years of Federal law enforcement experience. OI Special Agents previously have served at various Federal agencies, including other Federal law enforcement agencies such as the Bureau of Alcohol, Tobacco, Firearms and Explosives, U.S. Department of Labor, U.S. Department of Energy, the Naval Criminal Investigative Service, the U.S. Air Force Office of Special Investigations, the Federal Bureau of Investigation, the U.S. Secret Service, U.S. Customs and Border Protection, U.S. Drug Enforcement Administration, and various Offices of Inspectors General.

OI conducts and plans investigations of allegations of wrongdoing to determine willful and deliberate actions in violations of NRC regulations and criminal statutes. OI conducts investigations in accordance with the Quality Standards for Investigations established by the Council of Inspectors General on Integrity and Efficiency.

OI develops and implements policies, procedures, and quality control standards for investigations of licensees, applicants, and their contractors or vendors. OI conducts and supervises investigations of allegations of wrongdoing by persons or entities within NRC jurisdiction and maintains proactive investigative efforts and liaison with other Federal, State, and local law enforcement officials.

DIRECTOR AND
FIELD OFFICE REVIEW VISITS

The Office of Investigations' (OI) Director or Deputy Director annually visits each of the OI field offices that are co-located with the four NRC regional offices. During these visits, OI senior management places particular emphasis on enhancing effective communication among OI staff and internal stakeholders. The Director's visits include individual meetings with each OI employee to discuss a variety of subjects and to effectively address any concerns or questions. Additionally, OI Headquarters' investigation and support staff may accompany the Office Director and Deputy Director during visits to OI field offices, which provide opportunities for effective knowledge transfer and increased operational and programmatic awareness. These visits facilitate, encourage, and demonstrate open exchanges of ideas and expressions of differing views between OI senior management and its field office staff, as well as between OI regional senior management.

Field Office Review Visits (FORVs) are annual self-assessments conducted of each OI field office to support the goal of continuous improvement of OI's national investigations program. OI FORVs assess three major focus areas: operations, management, and administration.

Each FORV includes a meeting of field office staff to discuss current OI Headquarters initiatives and activities, policy and procedural focus, and special or regional items of interest. During the self-assessments, OI personnel also are interviewed to obtain timely feedback about operational or other concerns and any issues of particular concern to the employee. Additionally, the FORV team meets with internal stakeholders, the Regional/Deputy Regional Administrator, the Regional Counsel, the Enforcement Coordinator, the Office Allegation Coordinator, and any other regional staff deemed appropriate. These meetings are designed to solicit stakeholder input on the effectiveness of OI's support and ways to improve the quality, effectiveness, and efficiency of OI's performance.

At the conclusion of the FORVs, exit briefings are conducted with the OI SAICs and OI staff to discuss the findings and recommendations of the FORV team. A final OI senior management and OI Headquarters staff review of the FORV teams' findings is conducted to identify and implement best practices with a view toward continuous program improvement and investigative excellence.

CASES

Case Inventory*

Figure 1 shows the OI case inventory, which includes all Investigations and Assists to Staff conducted during the fiscal years indicated. Assists to Staff involve matters of regulatory concern for which the staff has requested OI's investigative expertise, but which may not involve specific indications of wrongdoing. The total case inventory in FY 2012 was 294. The total includes 245 investigations, 106 of which were carried over from FY 2011. Also included are 49 Assists to Staff, 12 of which were carried over from FY 2011.

Figure 1 Case Inventory

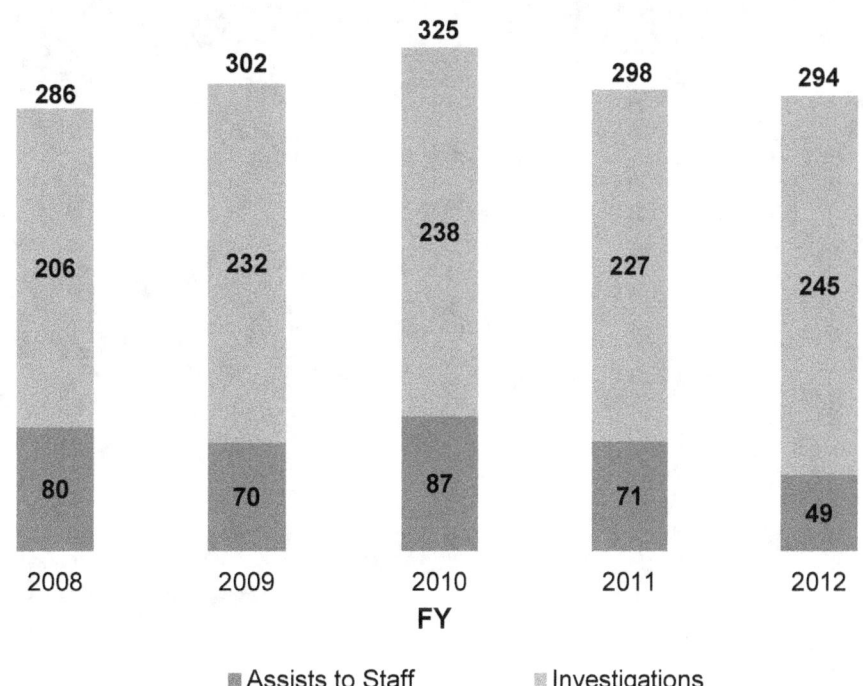

■ Assists to Staff   ▧ Investigations

* Cases carried over from previous year, plus cases opened in current

The total number of cases in the OI inventory during FY 2012 was 294, a 1 percent decrease from 298 in FY 2011.

CASES OPENED

Table 1 shows the number of cases opened by category during FY 2008 through FY 2012. In FY 2012, there was a 12 percent decrease in total cases opened from FY 2011. There was a 20 percent increase in the number of suspected Material False Statement investigations and a 6 percent increase in violations of other NRC Regulatory Requirements. In FY 2012, the number of Discrimination investigations decreased by 9 percent, and the number of Assists to Staff cases decreased by 40 percent. OI opened 176 cases in FY 2012 in the categories listed below:

Table 1 Cases Opened by Category

Category	FY 2008	FY 2009	FY 2010	FY 2011	FY 2012
Total	222	206	228	199	176
Material False Statements	21	23	21	15	18
Violations of Other NRC Regulatory Requirements	97	86	79	69	73
Discrimination	32	30	46	53	48
Assists to Staff	72	67	82	62	37

Note: Out of the 176 cases opened in FY 2012, 10 percent were comprised of Material False Statement investigations, 42 percent were Violations of Other NRC Regulatory Requirements, 27 percent were Discrimination, and 21 percent were Assists to Staff.

The graph in Figure 2 shows the distribution of cases opened during FY 2008 through FY 2012 for the Reactor and Materials programs. From FY 2011 to FY 2012, the overall number of Reactor cases decreased 17 percent with a 7 percent decrease in Reactor investigations and a 45 percent decrease in Reactor-related Assists to Staff. Materials cases increased overall by 18 percent with a 45 percent increase in Materials investigations and a 21 percent decrease in Materials-related Assists to Staff.

Figure 2 Cases Opened by Reactor/Materials

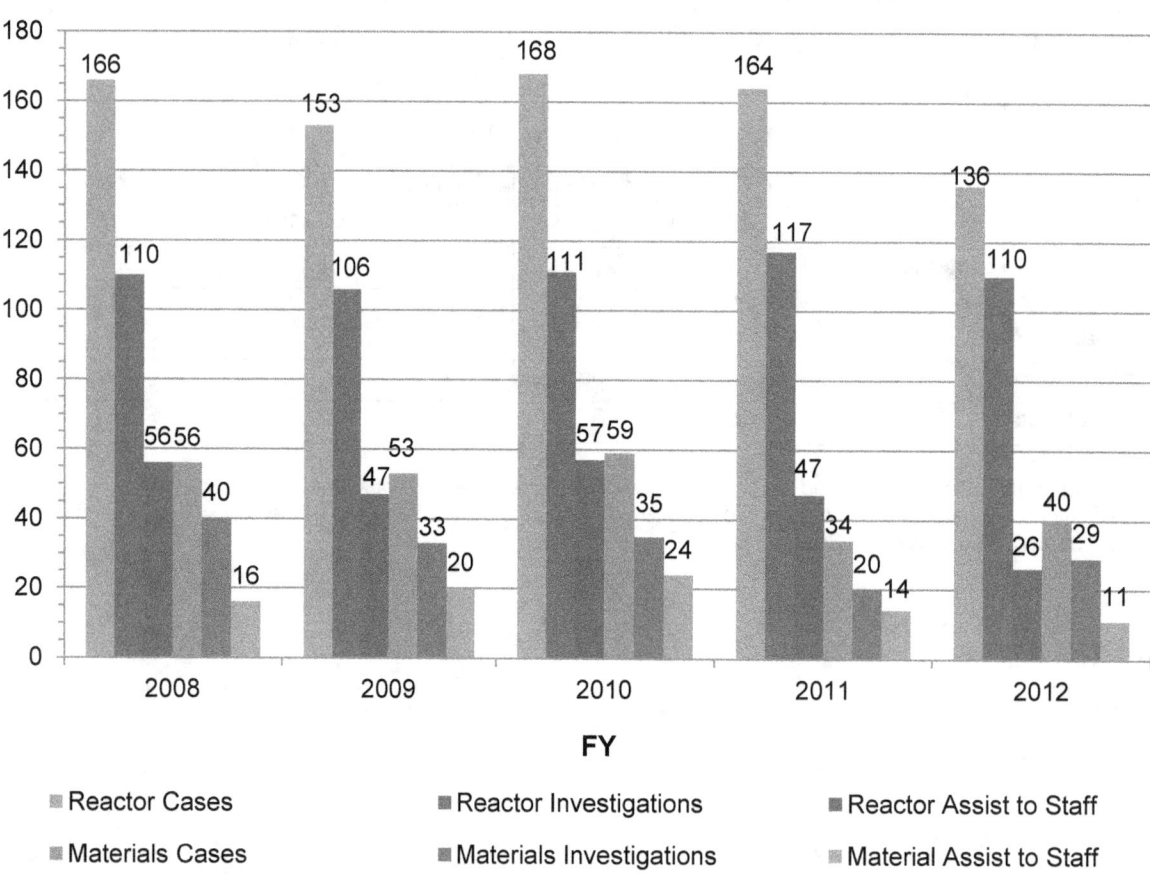

■ Reactor Cases   ■ Reactor Investigations   ■ Reactor Assist to Staff

■ Materials Cases   ■ Materials Investigations   ■ Material Assist to Staff

CASES CLOSED

Table 2 shows the number of cases closed by category during FY 2008 through 2012. The total cases closed during FY 2012 represent a 1 percent decrease from the number closed in FY 2011. There was a 17 percent increase of Material False Statement investigations and a 17 percent decrease of investigations involving Violations of Other NRC Regulatory Requirements. Discrimination investigations increased 70 percent, and Assists to Staff decreased 24 percent. OI closed 178 cases in FY 2012 in the categories listed below:

Table 2 Cases Closed by Category

Category	FY 2008	FY 2009	FY 2010	FY 2011	FY 2012
Total	191	205	226	180	178
Material False Statements	12	23	21	12	14
Violations of Other NRC Regulatory Requirements	77	90	85	76	63
Discrimination	25	27	42	33	56
Assists to Staff	77	65	78	59	45

Note: Out of the 178 cases closed in FY 2012, 8 percent were comprised of Material False Statement investigations, 35 percent were Violations of other NRC Regulatory Requirements, 32 percent were Discrimination, and 25 percent were Assists to Staff.

The graph in Figure 3 shows the cases closed from FY 2008 through FY 2012 for the Reactor and Materials programs. From FY 2011 to FY 2012, the overall Reactor-related cases increased 1 percent, accompanied by a 20 percent increase in Reactor Investigations and a 36 percent decrease in Reactor-related Assists to Staff. Materials cases decreased overall, accompanied by a 27 percent decrease in Materials investigations. Materials-related Assists to Staff increased by 7 percent during this period.

Figure 3 Cases Closed by Reactor/Materials

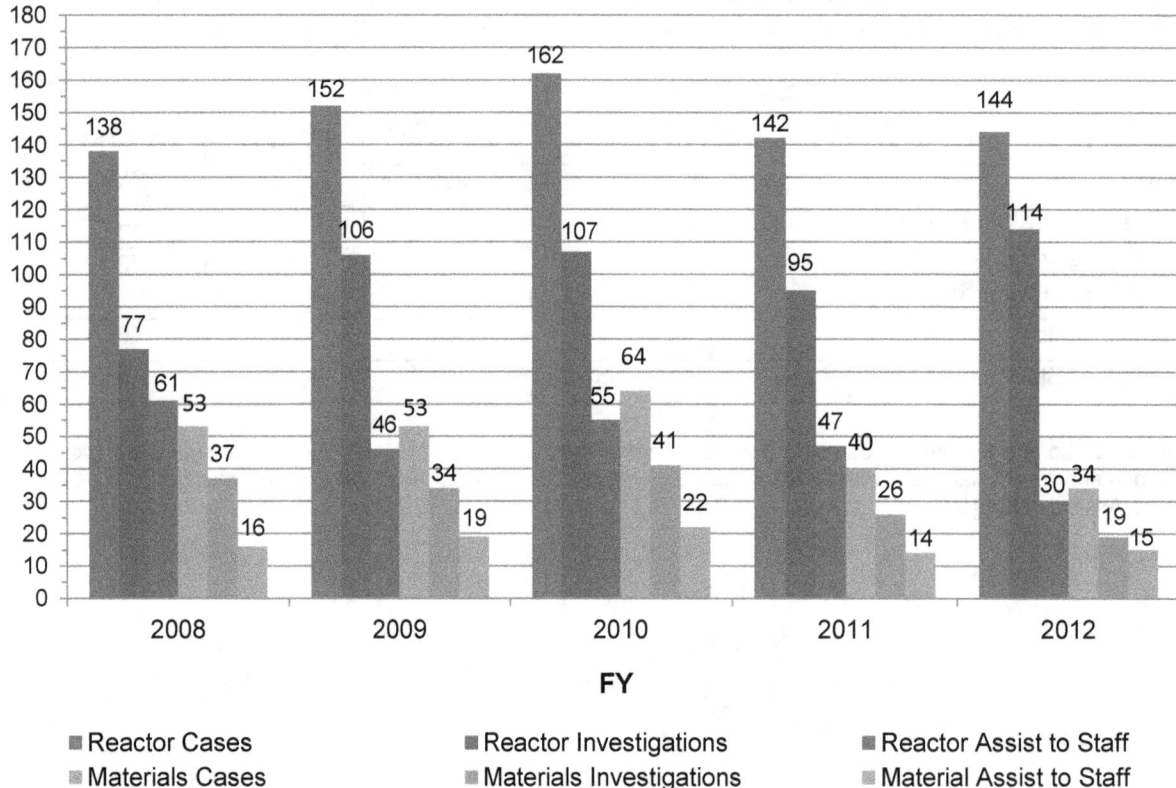

Of the 178 cases closed in FY 2012:

- 37 investigations were closed after OI substantiated willfulness on one or more of the allegations of wrongdoing

- 94 investigations were closed after OI investigations did not substantiate willful wrongdoing

- 2 investigations were closed administratively

- 45 of the total number of cases closed were Assists to the NRC Staff

MANAGEMENT OF CASES

The total case-specific staff hours (civil/criminal investigations) in Figure 4 shows an increase from FY 2011 to FY 2012 (from 31,000 to 32,000 investigative hours). In FY 2012, case activities (planning, field work, and analyzing evidence) stayed the same while hours toward case administration (Freedom of Information Act (FOIA) and other miscellaneous activities) increased.

Figure 4 Case-Specific Staff Hours

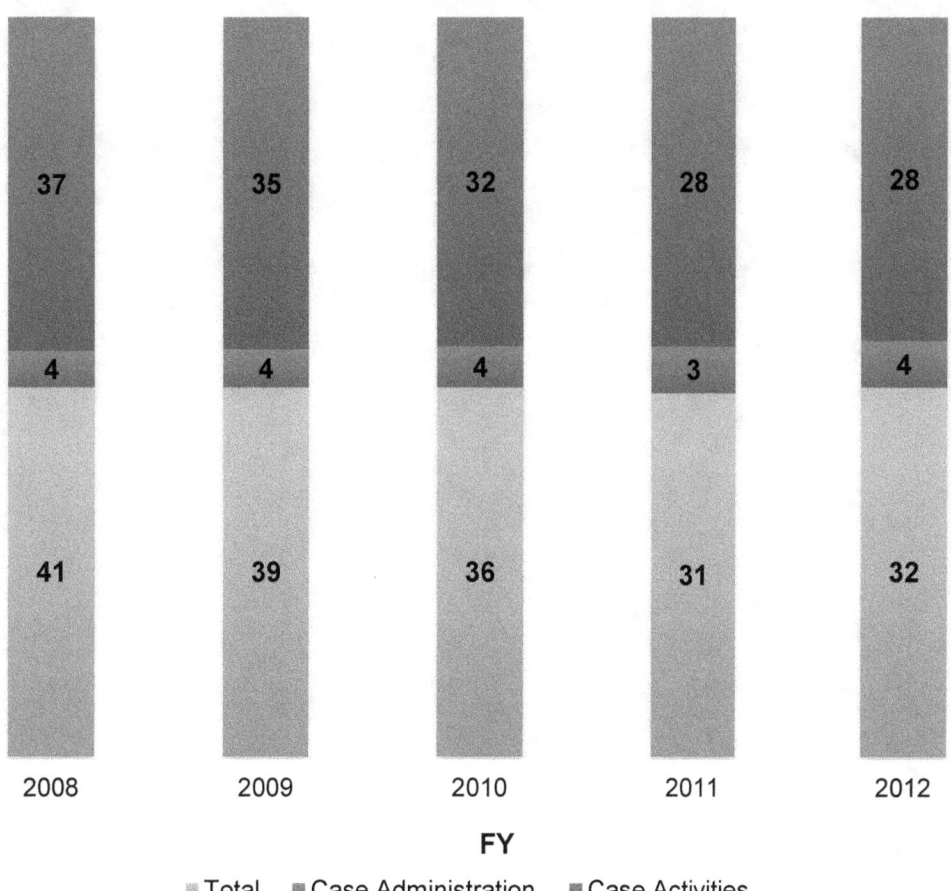

SIGNIFICANT INVESTIGATIONS

Arkansas Nuclear One

An OI investigation substantiated that four security officers were deliberately inattentive at their stations. This investigation was initiated after the NRC received photographs of three officers in an apparent inattentive state. All four security officers were acting as armed responders during the time they were inattentive, and they were found to be in deliberate violation of failing to remain in continuous contact with the station. This investigation was referred to the U.S. Department of Justice for prosecution consideration. On October 22, 2012, the NRC issued a Severity Level IV violation to the licensee.

Texas Gamma Ray

An OI investigation substantiated that a radiographer deliberately failed to follow increased controls for radiographic operations, and that the Radiation Safety Officer (RSO) willfully failed to implement the controls. The investigation involved a radiographer who knowingly stored the radiographic sources in an unauthorized location that was not licensed by the NRC. The radiographer made no attempt to facilitate the appropriate arrangements to store the source properly. The OI investigation determined that the RSO willfully failed to enforce the controls, as the RSO was aware of the unauthorized storage of the source and failed to intervene to correct the situation. This investigation was referred to the U.S. Department of Justice for prosecution consideration.

On September 9, 2012, the NRC issued a Confirmatory Order to the individual. Previously, on May 15, 2012, the NRC issued a Confirmatory Order to the company.

Lapeer County Surgery Center

An OI investigation substantiated that an office manager for the Lapeer County Surgery Center (LCSC), a licensee, willfully scheduled and allowed the performance of radioactive medical procedures without having a radiation safety officer (RSO) on staff or affiliated with the LCSC license to conduct such procedures. Specifically, between November 2009 and December 2012, LCSC conducted eight prostate seed implant procedures after the LCSC RSO left over a salary dispute. NRC regulations specify that a licensee must have an RSO, and must notify the NRC within 30 days when the RSO leaves or when a different RSO is named. Although the NRC staff did not find a violation of these specific requirements, it did find that LCSC committed a nonwillful violation of Title 10 of the *Code of Federal Regulations* (10 CFR) 35.24(e) by failing to describe in writing the authority, duties, and responsibilities of the RSO. LCSC voluntarily canceled its own NRC license to perform radioactive medical procedures. This investigation was referred to the U.S. Department of Justice for prosecution consideration.

On October 19, 2012, the NRC issued a Severity Level IV violation to the hospital and a closeout letter to the individual.

Columbia Generating Station

An OI investigation substantiated that a former Condenser Replacement Project Superintendent (CRPS) at Energy's Northwest Columbia Generating Station submitted synthetic urine during a random Fitness for Duty examination. The investigation revealed that the former CRPS falsified a Federal Drug Testing Custody and Control Form, wherein he certified that his submitted urine specimen was "not adulterated in any manner." However, the CRPS admitted to OI that in addition to taking prescription medication, he purchased and used a synthetic urine kit, which he brought to the plant drug testing facility. This investigation was referred to the U.S. Department of Justice for prosecution consideration and remained under NRC regulatory review.

Watts Bar Nuclear Plant

An OI investigation substantiated that two subcontractor employees (an electrician and a foreman) at the Tennessee Valley Authority (TVA) Watts Bar Unit 2 Nuclear Plant deliberately falsified work order packages for primary containment penetrations. The electrician admitted to OI that he deliberately falsified micrometer readings identified in the work order packages and falsely annotated on the work orders that micrometer readings had been performed for cables in these penetrations, when the micrometer readings had not been completed. The foreman also falsely attested that a work order review, field walkdown, review of craft documentation, and the scope of work all had been completed. TVA opted to request alternative dispute resolution (ADR) and a post-investigation ADR settlement was reached, in which TVA agreed to complete broader comprehensive corrective actions. Based on evidence obtained during the investigation, NRC staff issued a Confirmatory Order to TVA on June 18, 2012. The NRC issued closeout letters to two individuals on July 16, 2012.

Paducah Gaseous Diffusion Plant

An OI investigation substantiated that a Technical Security Specialist at the United States Enrichment Corporation (USEC), Paducah Gaseous Diffusion Plant (PGDP), deliberately falsified security container logs at PGDP. The inaccurate inventories led to discrepancies in the accountability of security locks. The OI investigation determined that the Technical Security Specialist altered the security logs and inventories in an attempt to cover up false entries that the individual made. In addition, it was noted that padlocks were not located within a security container. However, these locks were previously documented to be stored in the locked security container during audits that the Technical Security Specialist conducted. This investigation was referred to the U.S. Department of Justice for prosecution consideration and remained under NRC regulatory review.

NRC FORM 335 (12-2010) NRCMD 3.7	U.S. NUCLEAR REGULATORY COMMISSION	1. REPORT NUMBER (Assigned by NRC, Add Vol., Supp., Rev., and Addendum Numbers, if any.)
	BIBLIOGRAPHIC DATA SHEET *(See instructions on the reverse)*	NUREG 1830, Vol. 9

2. TITLE AND SUBTITLE		3. DATE REPORT PUBLISHED	
Office of Investigations Annual Report FY 2012		MONTH	YEAR
		February	2013
		4. FIN OR GRANT NUMBER	

5. AUTHOR(S)	6. TYPE OF REPORT
	Annual
	7. PERIOD COVERED (Inclusive Dates)
	10/01/2011 to 09/30/2012

8. PERFORMING ORGANIZATION - NAME AND ADDRESS (If NRC, provide Division, Office or Region, U. S. Nuclear Regulatory Commission, and mailing address; if contractor, provide name and mailing address.)

Office of Investigations
U.S. Nuclear Regulatory Commission
Washington, DC 20555-0001

9. SPONSORING ORGANIZATION - NAME AND ADDRESS (If NRC, type "Same as above", if contractor, provide NRC Division, Office or Region, U. S. Nuclear Regulatory Commission, and mailing address.)

Same

10. SUPPLEMENTARY NOTES

11. ABSTRACT (200 words or less)

This report describes Office of Investigations case activities during FY 2012.

12. KEY WORDS/DESCRIPTORS (List words or phrases that will assist researchers in locating the report.)	13. AVAILABILITY STATEMENT
Office of Investigations FY 2012 Annual	unlimited
	14. SECURITY CLASSIFICATION
	(This Page) unclassified
	(This Report) unclassified
	15. NUMBER OF PAGES
	16. PRICE

NRC FORM 335 (12-2010)

Printed
on recycled
paper

Federal Recycling Program

NUREG-1830, Vol. 9

February 2013

www.ingramcontent.com/pod-product-compliance
Lightning Source LLC
Chambersburg PA
CBHW081613200526

45167CB00019B/3476

*9 7 8 1 4 9 9 6 2 9 5 7 6 *